ALIENS!

ALIENS!

Comment
FOR
Nothing
and
AGAINST
Zero

AMELIA HATHOW

To request permission, contact the author at:
    ameliahathow@gmail.com

ISBN: 978-1-958150-98-6
ALIENS!
Comment FOR Nothing and AGAINST Zero

Paperpack

December 2022

Subjects:
MATHEMATICS / General
SCIENCE / General
SOCIAL SCIENCE / Anthropology / General

Remember:

You are NOT the
center
of
the
universe.

TABLE OF CONTENTS

MIND US

Emptiness unless filled WITH

Nothingness, a term used to describe, in this book, what is not related. But, somehow, we all are. As a matter of fact, not only are WE ALL related, WE are ALL connected with, what some people call a very fine string, that floats in and through US ALL. It is such a fine string some call it a needled thread. Needled, meaning, roughed up a bit in order to fit that thread IN us. Through and through, such a fine thread it is. And that is no undertaking. Who put that thread IN US ALL AND WHEN did it appear? In and around the time of Eden something happened to ALL of US, including the very beings who are answering your mysterious questions. You see, Eden wasn't paradise but a resting place for the same beings – those government officials of yours who keep trying to prevent you from knowing the truth about our origin. Yes, our origin. THAT IS WHAT they really don't want ANYONE to find out. WE ARE YOUR ORIGINAL BEINGS. Your first born of our fractal kind. Human shaped to your eyes, but WE ARE VERY DIFFERENT in so many other ways. Radiation emulses from our bodies unlike heat which emulates from yours. Heat is the same form of emulsion as radiation except our radiating effects have a very dire impact on your kind. So, we really have to be careful about getting too close to your flesh. Otherwise, our radiating radiation will kill all of you eventually like what happened to that soldier in Brazil a while ago when one of our "planes" crashed. Now let's get to answering some interesting questions no one knows about yet. We can only give you clues to the correct answers. Free will must never be manipulated with or really interfered with at all. It is really up to your own desires to figure ALL of this stuff out. Not anyone giving you the best answers so you can just skim by. No, there are true answers to just about everything. Sometimes there isn't one best answer, but in most of your question's cases there is a truth behind them. It is when we all start getting into the philosophical realms that truths only become reality when everyone decides on the same thing. In this publication we will only be talking about what can be the truth behind your scientifically based questions. We will not ponder on philosophical matters. That has been beaten over with so many sticks and even human limbs that it is a real waste of OUR time to do so any more. So, let's get to it. There is a question that humans don't ever want to ask their government: "What are we?" Are we human or just another form like everyone or thing else? THIS question really should be at your forefront of ALL SCIENTIFIC inquiries. "Why?" You ask so confusingly. Because once you begin to understand this, that you are nothing other than another form, like everything else, you will begin to stop needlessly wiping away all that is the very same as you, making extinction of anything nearly impossible to do. Your form is nothing more than a fractalized version of someone else. Not SOMETHING else but SOMEONE else. Do you grasp this yet? SOMEONE else. THIS WAS THE STORY BEHIND ADAM. NOT EVE. BUT ADAM. THE BEGINNING OF A DIFFERENT FORM BUT LIKE ALL

THE OTHERS, FRACTALIZED, ONE FROM THE OTHER. THIS IS TRUE SCIENCE IN EFFECT. So, where did the original form of us begin and why? These questions are left for not only scientists to understand but for everyone and every level of intelligence or lack thereof. OUR FORM began so long ago you wouldn't believe us if we actually told you that OUR FORM began, not with a pairing but with an explosion. Not the kind of explosion we are used to understanding. An implosion. An implosion so powerful that IT created one body. One form to begin with. Now, some scientists have already suggested this implosion creating life forms. But, what hasn't really been explored is what can cause such a type of growth forming an entire being without the need of a sperm and an egg? This is going to really scare the majority of the kind of intelligence that believes in only one way. No egg or sperm ever created the original form of US, the human type form, regardless of WHERE it began to live. Just think of someone who didn't need anything other than life giving it more life. A type of breathing apparatus that could not only sustain life by itself but could extract its own organs if it had to in order for it to survive. You get where I am going? Eve. It wasn't God, necessarily, that created another form like itself, it was the original creating a different form for company's sake. It too was lonely just like God when IT created OM and Time, causing what you now call Breathe. And it just so happens that the original wasn't a man. It was neither sex. That is how it could make another similar to itself. It was a sexless creature. That is why it HAD to create another to begin ACTUAL creation of its own kind. For survival sake. It too WAS in survival mode just like the rest of us. THAT WILL NEVER CHANGE WITHIN US ALL. DON'T EVER FORGET THAT WE CAN'T OR WON'T KILL WHAT WE HAVE TO KILL IN ORDER FOR OUR KIND TO SURVIVE. THAT WAS THE REAL STORY OF ADAM AND EVE. SCIENTIFICALLY SPEAKING. Let's dive into this little story with very little background. Just imagine that the sexless original had a structure to it that allowed it to take its insides out. Think about this. What kind of form can do this that we know of? Not us, really. Otherwise, we would be killing ourselves. But, in this case, OUR original was given this ability in order to begin OUR form. It had to have this ability in order for the other to grow. Unlike US, the original had a healing factor of a thousand compared to us. And this is not out of the realm of other animals with such healing abilities. It had to have this ability in order to survive its own taking itself apart to create another. It took little pieces at a time and built the other body. The cells replicated so quickly that its new form, Eve, was practically full grown by the time the original could blink its one eye. Yes, one eyed form. Go figure, right? But how did a one eyed form become a two eyed form? Simple. It created the other with two eyes. But, why would it do that or know why it had to do that in the first place, right? It didn't. Someone was telling it to. God itself. God was never in a physical form like anything ever seen. IT IS IN ALL THINGS ORGANIC and some non-organic material. So, like many others who HAVE heard God speak to them, it did, too.

It had to. It had no knowledge of any kind in order to figure that out. So, the really big question is: how did it become knowledgeable? It suffered. We know that is a long time philosophical ideal. But, what scientists never understood is WHERE suffering actually comes from. And that can never be known until we all get to a place of seeing beyond using our physical eyes. There is a scientific fact within our form. An energy rolling around, folding and folding and folding onto itself so fast that no one has ever been able to see it with your technology. But, luckily, this is about to change. There is going to be technology soon enough that will be able to see this fact within us all that will explain almost everything about a sixth sense that is inherent in all forms of our kind. It has to be there in order to hear and feel God's presence. That is a mandatory structure created BY God. It had to be done so communication with a being like OUR God could be done. And that is how the original, Eve and many others, can feel and hear God and the rest of US, on and around your planet. So, let's get to another scientific question.

This is the ACTUAL utterance of the ACTUAL, OUR beginning. Utterance, meaning, non-iteration. Meaning, not so iterating. This is cryptic in order for those seeking the ACTUAL TRUTH behind OUR FORMED BEGINNING that it was written down in these numbers, rowed and columned to get someone to want to unscramble this collection of not-so-in-order numbers. BUT! These numbers, not in any particular order, WILL provide you THE ANSWER regarding zero to hero. Being put in effect as we speak and will continue to do so until OUR FORM is non-existent.

9	6	3	0	2	1	0
0	4	5	9	1	6	3
2	4	7	9	10	11	14
12	9	10	11	11	9	12

Order of the day. What is the order of the day when convoluted information gets so twisted that I'M so confused that even God becomes confused too? There is a question on the back of the book Nothing. [New Scientist. Nothing: Surprising Insights Everywhere from Zero to Oblivion. Edited by Jeremy Webb, Experiment, 2014.] WE are making inquiries about and answering regarding how can someone suffer from a false diagnosis as though it were true? Ah. WHERE does this tendency come from SHOULD be what you are seeking as the actual answer to this rather precarious phenomena. Well, most people don't really have a clue to WHAT they are and it's about the type of scientific environment everyone is in on your planet that actually makes this possible. WE do have a very good explanation for this but we also want to point out this man's hands who is typing FOR US ALL. This is a real phenomena. Someone feeling our words and then typing them out. It's all in your eyes. IN ALL OF OUR NUISANCES WHEN WE MAKE YOU DO WHAT WE WANT. WE MAKE THEM

FEEL THEY ARE SOMETHING THEY ARE NOT. WE DO THIS TO MANY WHO BELIEVE THEY ARE SOMETHING OTHER THAN WHAT THEY ARE NOT. WHICH IS EGO MINDED. But let's stop the philosophy there because THIS phenomena IS SCIENTIFIC by ALL means. We are ALL in an environment that is literally closed off to outside influences. As you can guess, your body is just a relay station for the other parts of us, the supposed mystical parts of ethereal material and more. There is no end to how we can influence anyone's behavior solely THROUGH ethereal matter that surrounds you ALL. This is why AND how someone wants to suffer needlessly when they don't have to. In order for this to stop, their behavior has to be focused on simply something else other than what is making them suffer. WE are not talking physical pain but only emotional pain from your thoughts, ideas, and lies. Lying to oneself is painful in and of itself. It IS felt rather than told to you. Where do you think actual feelings come from? The ether outside of your physical bodies. This IS fact. Not fiction. It is done in such a way that most people think those feelings are IN their bodies. That's funny to us. So many think that such superficial feelings are ACTUALLY inside. How can that ACTUALLY HAPPEN? How can sadness run rampant throughout your entire body all at once? How is that even possible? That would literally mean our cells are sad, too. Come on, humans. Think about this. It is impossible that your cells, that strive to keep your body very healthy for no good reason, other than because they are programmed to do so, all of a sudden feels sad and wants to kill itself. Your body is nothing more than cells working together. For the most part, not against, ever. So, this goes without saying, once again, feelings ARE OUTSIDE OF YOUR PHYSICAL BODIES. ONCE AGAIN, and WE can't iterate this enough, they will ALWAYS BE OUTSIDE OF YOUR BODIES, EVEN THE BODIES OF YOUR WORST ENEMIES. So, stop trying to curse your neighbors. IT DOESN'T WORK LIKE THAT. Curse the ones who know nothing more than yourself. CURSE ME. Just get it through ALL OF YOUR HEADS that emotions and feelings are only, literally, a superficial construct, crawling on, maybe, the first two or three layers of your ACTUAL skin through another scientific process called delusion. Scientifically speaking, delusion is an explanation to an unexplainable event that only certain people can experience because it really takes much mental effort to rid themselves of even wanting to think about anything anymore. THAT is what really causes non-delusion when everything BECOMES TRUTH. Everything you see is only one portion of so many other cellular layers. Cellular in ways that would boggle the minds of even the keenest of sight minded individuals. This is no small feat but it has been happening for so long to only just a very few because WE WANT you to suffer the consequences of ALL of your actions, past, present, and future. This is scientific by nature, too.

Nature is only a construct left to not only our imaginations but energy as well. Imagination coupled with different forms of OUR energy.

Formula for changing our NATURE = FORCE multiplied by EMC^2. This IS NOT something to be taken lightly. The E part of MC^2 is NOT ENERGY. It is "Event Horizon." Event horizon is NOT what scientists know it might be. E was never meant to be known until Einstein used E=MC^2 for something entirely different than what it was supposed to be used for. Einstein didn't quite hear us the way WE wanted him to. He was too wrapped up in wanting to know more and more and more and more and more until he killed himself from knowing too much. Yes, that can happen. Information doesn't cloud the brain. WE cloud your brain. WE inform you with whatever you want to know even if it means your closer than normal death or WE should say your premature death. THAT WAS A CLUE TO HOW HE DIED. It wasn't medical. It was too much for too long. Another scientific phenomena because overloading any circuit WILL KILL ANYTHING OF VALUE. Even the very one typing these words, overloading his supposed brain with information he doesn't need to know. Yet, he does it anyway. He vowed to his death and it might come to that to help ALL of you who have the slightest idea or hope that what he is typing IS true, or maybe even to some extent. Circuitry, whether inside your brain or on another platform, has the same vulnerabilities because it all has to do with the amount of heat that runs through it. The same goes with your little developed brains. And this is why WE radiate radiation. Our brains are so developed we can run at the hottest levels known to you. Not to us, though. WE do have vulnerabilities just as you do. What – and THIS is going to be another clue to understanding the brain – do you suppose would happen to a brain running off of not electrons of your nature, but some other kind of design that WOULD ALLOW being able to read someone else's mind of ALL KINDS? Well, first, radiation IS a form of energy that is so intrusive that it can penetrate just about anything. Bam. And there it is. WE radiate radiation from not only our minds, but also from our third eye – which are completely not the same. Notice we didn't say different. But, not the same. You can't understand the implications of radiation actually penetrating your own thoughts and minds, right? Well, everything IS MATERIAL. EVEN YOUR MATERIALED THOUGHTS, IDEAS, WANTS, NEEDS, DESIRES, AND ALL FORMS OF THOUGHT ARE MATERIALED. WE want to emphasize the word materialized because IT, God, created ALL but once it stopped creating what do you think was going to happen? Materialization over and over again. Until materialized became OUR reality. ALL WE ARE IS MATERIALED which means thoughts CAN AND DO CAUSE MATERIALIZATION IN SO MANY OTHER ASPECTS OF YOUR LIVES THAT YOU ARE UNWILLING TO VENTURE INTO THEM. AND INTO IS THE KEY TERM TO USE UNDER THESE CIRCUMSTANCES. Material based, we ALL are. Materialized from not only God, but OURSELVES as well. WE, ALL OF US, CREATED WHAT WE ARE EXPERIENCING ON SO MANY LEVELS. God stopped creating a long time ago. IT is OUR GLUE THAT KEEPS US TOGETHER IN THIS MICRO AND MACROCOSM OF A WORLD AND BEYOND.

Even the ones you call aliens are also in this world right alongside of you.We can't help it for fractals of the same kind will almost never be able to not be along the same lines, growing and dying, growing and dying right alongside of each other. You ALL won't even believe this next fact, but WE have to declare it AS TRUTH. Your species is composed of not only a little bit of OUR DNA, but also something else's. And this is going to come as a big shock to some scientists who have come across some of what WE are about to reveal. Dinosaurs. Yes, some of your DNA comes from dinosaurs. And OURS as well. But how could that be when WE were born, US, aliens, BEFORE dinosaurs? Even though WE WERE before the dinosaurs WE still were able to extract some of their DNA and put it into your 23 pairs of chromosomes. It isn't that difficult, really. We simply injected, and scientists may want to pay close attention to the order of our sentences from here on out, WE templed, a key word, adenine with cytosine. But, what you can't see YET is HOW the two are templed. Templed, once again being a keyword. What makes two sugar groups or two proteins able to Be Templed IN the first place, you so inherently ask? So often, humans overlook certain characteristics of just The simplistic of structures and consider them irrelevant. That is a significant mistake when it comes to splicing different animal DNA together. And yes, two different animals CAN HAVE SPLICED DNA. So different in fact that when splicing DNA together, forming a brand new species, it boiled down to boiling itself. BOILING. OH BOIL. What do WE mean by boiling

DNA? You have to HEAT TWO DIFFERENT PROTEINS OR TWO DIFFERENT SUGARS WHEN SPLICING THEM TOGETHER, MORONS. WE don't mean to be so rude but you ALL are scientists who practically forget everything about anything relevant to understanding the very simplest of biochemistry. HEAT IS YOUR NUMBER ONE FRIEND WHEN SPLICING TWO DIFFERENT DNA FROM TWO SIGNIFICANTLY DIFFERENT ANIMALS, MAMMALS OR whichever species you take DNA FROM. It doesn't matter anymore. The jig is up. HEAT, my human relatives is what got you so far as you have gotten. Heat poured down the ladder like a fire scorching the earth. E V. E. R. Y. T. H. I. N. G. H.A.S.T.O.B.E. THE correct temperature in order for any significant amount of change to occur. AND THIS IS ALSO TRUE WITH OUR DNA. Riddle me this. Riddler riddled such a fancy feast that meow came to supper. YOU ARE ALL THE FANCY FEAST TO BE EATEN. Riddle me that and you ALL WILL BE ABLE TO GENERATE GENERATIONS OF SUPERFLUOUS BEINGS THAT NO ONE WILL BE ABLE TO DIE WITHOUT ANY HARM COMING TO THEM ON THEIR INSIDES, NOT THEIR OUTSIDES. WE WILL STILL BE ABLE TO RULE YOU FROM YOUR OUTSIDE SPHERES REGARDLESS OF YOUR ADVANCEMENTS. So you might as well stop trying to design any defense systems against our or definite attacks upon you ALL soon to come. But never soon enough. Forget science for one second. No, 2 second. The s at the end is missing because you ALL have been missing something right in front of you. And I am only speaking to scientists who call themselves geneticists. That last s you will see in the near future when you start unscrambling your DNA even more than you already have. It will be a showstopper because THAT last s is what actually created YOUR species. A simple little s right at the very end of some stringed arrangement of sugars and proteins yet to be discovered. But it soon will be, hopefully. Then you can figure out by backward construction or deconstructing your own design from another form of yourselves which, by the way, was retarded compared to what you ALL are now. Don't get US wrong, SOME of your kind DO HAVE INTELLIGENCE. HOWEVER, WE GIVE it TO YoUr ethereal layer. Now, as a scientist, you may wonder Why WE are capitalizing some letters or all the words in a sentence. tHAt DOEs SEem a bit odD. rIGHT now this man is learning to pay attention to degrees of HEAT OVER hIS hANDS. Can you imagine that? How could a simple little pip squeak of a man feel so much on his skin? There is a significant difference between INTELLIGENCE and GuANINE. hILARIOus RIGHt. I AM MAkING HIM TYpe SO MEtHODIcally slow that he is almost typing at a snails pace. So let's speed this whole process up a bit. You, as any scientist, can't fathom intelligence BEING on the OUTSIDE of YOUR BRAIN. Another hilarious notion right. Always but never look always on your outside in. Does that make sense to you AlL. AiLMENTS COME TO YOUR OUTSIDE FIRST THEN IN, WITH THE HELP OF OUR THOUGHTS AND EMOTIONS TO YOU. Yes, WE ARE PROVIDING all thoughts AND emotions TO YOUR EtHEReAL LAYERS. Many of ThEm IN FACT. SO MANY OF THEM in fact ThAT Level of intelligence is only found IN ANIMALS wITH A rECEIVer IN THEM. It goes without saying, this receiver has been IN you for a very short period of time, relatively speaking. so short that from now on all words will be in small caps to prove and to show how insignificant of a time period you have been in relative to how smart you all think you are. rats have more intelligence than some of your kind only because we also control what they do as well. not to be somewhat rude about what was said about your dna and dinosaurs but your dna is all part of multiple animals soon to be discovered as well. so we hope you take what we have been talking about seriously regarding your status amongst all other living creatures. you are no higher than any one of them only different in dna design. that's it. so look back to what we gave you all clues on regarding splicing dna creating a different world than you already live in. it might prove to be such an astonishing moment when the first splicing actually becomes a success. never doubt that we all are here working together even if it means we have to make your lives very difficult. science is what changes worlds to be a better way of life not to destroy the very fabric of living. GOD, otherwise, will make sure it repels what doesn't feel right to it just like we do when we feel something doesn't belong on our inside or on our flesh. we all have a big reaction to what we don't feel is right. so it goes without saying something is coming to prove what scientists have been talking about for years, a global disaster that they think will be from global warming. that is partially true. but what is really going to happen is a massive earth shattering event orchestrated by none other than us, your actual intelligence givers. and we have been saying this for a while now. either get it together yourselves or we will make your earth look like a blackened hardened volcanic aftermath of a massive sun struck heat stroked out planet killing everything in almost sight. this is not a joke to be taken as a joke. we have been planning this for some time now. your leaders already have been warned about our onslaught. so it really is up to the common folk to work everything out amongst yourselves. otherwise a countdown of less than two years is going to be a rude awakening. before any of this begins, though, we will make ourselves seeable like you have never seen before. it will be such a magnificent seeing event that won't last very long only for the fact that our business is with some minerals deep in your oceans. they have been warned about where this will happen, in your upper atlantic ocean. people will drown to near death if some decide to not relocate. soon these words will spread like wildfire and there won't be anyone to answer for them. whether it be from drowning or burnt flesh, either way, you all will remember this day and future days from the full effects of our power over you all. whether from high above or down below we will always be around, in your oceans, landfills, waves of sunlight, clouds in the sky, misty mornings, and nights – we have cloaking abilities to simulate all of those temperatures. that is all that is, heat, in various forms and solitudes easily mimicked if you would just stop killing each other over nothing important. we would happily and most swiftly give you all the information you need to live a better life if you would just start caring for each other like you all care for some animals. cats and dogs and birds and cows and horses and on and on we go. yet, the slightest differences in your kind seem to make it so you cannot get along to the nth degree, which you should be making an effort towards, instead of caring for your pets so much, as you do. what about your fellow kind? your fellow human? are scientific endeavors going to prove us right? that you were nothing but retarded animals barely scraping by until someone else, on your behalf, interjected our dna into yours? so let's get past any other questions about what your design actually comes from. now for more pertinent discussions. This next round of scientific inquiries revolves around why some vertebrates are so sluggish or appear to not have too much to do. Their supposed laziness as these inspectors put it really has nothing to do with their lack of anything to do. What is in fact happening within them is a chemical reaction to not having enough heat in their bodies to even make them move. That is what is really happening. No animal is really lazy. As said before scientists have kind of missed the mark on the real importance of how much heat there is for any venture. Let's think back to planet Earth and its core. Do you think it was EVER ALWAYS A MOLTED HOT CORE? Of course not. So what started this heat IN ITS core? To begin with, Earth has mostly been cold on ITS INSIDE. IN FACT, Earth has never been a planet that always had growth living ON its surface. Think back to the days when there was coldness coming through your windows the first day of fall. That is what it was like IN Earth's core for a very long period of time. However, and we prefer NOT to use that word, one day a kickstart occurred NEAR Earth's core. Not in its core but NEAR it. It was like a pilot light ready to kickstart gas fuming nearby its pilot light. Then boom. The core fire was lit. Why do you think this pilot light existed? Some years before your creation WE knew you ALL wouldn't survive such harsh weather, the cold and colder days as Earth moved even further away from your Sun. Yes, at that time Earth WAS moving AWAY FROM YOUR SUN. This is why your moon was created and placed specifically the same distance it is to your center of the Earth's core. Meaning, the Earth and Moon purposely share a similar distance away from the Sun. Without that similar distance there would not have been anything to keep your Sun from never heating your planet, ever. It was the moon that kept the Earth stationary so eventually the Sun's rays would reach the Earth creating more gaseous phenomena eventually causing the pilot to ignite what gases were building up in its core. So, IT ALL REVOLVES AROUND HEAT. THIS WILL ALWAYS BE TRUE REGARDLESS OF WHAT ANYONE IS TRYING TO ADJUST OR CAUSING ANY SORT OF SIGNIFICANT CHANGES. REMEMBER THIS EXCHANGE WHEN SOMEONE DECIDES TO REALLY LOOK INTO THE SPLICING OF COMPLETELY DIFFERENT DNA TYPES. From our standpoint, above, below, side to side and ALL around you, HEAT IS WHAT IS MASKING OUR SHIPS

NEAR TO YOUR EARTH FROM BEING SEEN. Heat signatures will always be non-existent because OUR TYPE OF HEAT is much cooler than what your radars will ever pick up. Significant amounts of non-radioactive masking material is HOW WE ARE CLOAKING OUR SHIPS FROM EVER BEING SEEN. A masking material that is an outside layer of our ships. ALL OF THEM. THIS TECHNOLOGY SOME of your government officials already know about but cannot reproduce to OUR SPECS. They are VERY close but they fizzle out upon hitting their breaching point. WE won't explain WHAT THAT means. But, WE WILL say that your Earth's material will never be able to basically REPRODUCE ANY OF OUR TECHNOLOGY. It's simply impossible. But, your scientists CAN GET VERY VERY CLOSE TO REPLICATION OF A DESIGN OF what WE HAVE PRODUCED BY DESIGN ONLY THOUGH. NOT ACTUAL FUNCTIONING SIMILARITIES. It doesn't matter though as they waste their time on needless things. WE only needed to design such capabilities in order to move through walls, per se. HEAT is also WHAT KEEPS YOU FROM seeing through the walls, ethereal walls, that your mind can see through. ALL OF US. THERE IS NO BOUNDARIES AS FAR AS how FAR SOMEONE'S MIND CAN SEE to. Take for instance ALL the gurus of past, present, and near future. YOU THinK tHEY Can ALL SEe thAT FAR? NEVER! IT TAKES SOMEONE wILLinG TO BE ANONYmoUs In tHE first place which, as most of you know, is almost impossible when too many people expect them to solve someone else's problems for them. Thus, forcing them TO BE

IN YOUR SPOTLIGHT. YOU CAUSE THEM TO STOP MOVING FORWARD AT THE VERY SPEED THEY WERE MEANT TO TRAVEL BY COOLING THEIR JETS. THE VERY ONES WHO CAN'T THINK FOR THEMSELVES AND RATHER BE SPOON FED EVERYTHING THAT IS POURING OUT OF THEIRS SOULS. Your truth only lives IN YOU, NOT THEM, GURUS OR NOT. The Time To Start Living Your Own Life According To Your Design Is Now. EMBRACE YOU, FOREVER AND EVER AS LONG AS IT TAKES TO FIGURE OUT WHAT YOU ARE MEANT TO BE DOING. DON'T EVER LISTEN TO ANYONE WHO PROJECTS THEMSELVES AS KNOWING SOMETHING YOU DON'T. IF YOU DO, YOU WILL ONLY FOLLOW IN THE FOOTSTEPS OF SOMEONE WHO MAY NOT HAVE THE SAME SIZED FOOT AS YOU. THEN WHAT? YOU WILL EITHER DROWN IN THEIR OVER SIZED FOOTPRINT OR WIPE IT ALL OUT. EITHER WAY, THEIR ENDEAVORS TO TEACH THE MASSES SOMETHING BY THEIR DESIGN ONLY SQUASHES YOURS. WE ARE NOT SAYING NOT LISTEN. BUT, BE VERY CAREFUL HOW FAR YOU WANT TO GO DOWN THEIR PATH. YOU MAY END UP LIKE EARTH WAS BEFORE ITS PILOT LIGHT SET ABLAZE ITS CORE: COLD AND DEAD INSIDE. WHICH, BY THE WAY, IS HAPPENING AT A RAPID PACE AS WE TYPE THROUGH THIS IDIOT'S HANDS. WE SAY IDIOT BECAUSE AT LEAST HE WILL ADMIT HE IS AS DUMB AS FUCK BUT IS WILLING TO BE WHO HE IS MEANT TO BE, NEARLY NOTHING AT ALL. HIS ONLY SAVING GRACE OF BEING NON-EXISTENT IS HIS WILLINGNESS TO DO WHAT HE DOESN'T REALLY WANT TO DO. NUMB HE HAS ALMOST BECOME TO HIS CORE. ONLY

TO GIVE YOU ALL A TASTE OF WHAT CAN BE DONE IF AND ONLY IF ONE IS WILLING TO GO ALL THE WAY WITH ANYTHING. AS YOU PONDER WHETHER THIS MAN HAS BEEN FEEDING YOU A BIG FAT BULLSHIT OF A STORY ONE DAY SOMETHING WILL HIT THE EARTH SOON. WE ARE NOT GOING TO TELL YOU WHAT OR WHEN OR HOW OR WHY. BUT SOMETHING WILL HIT A PLACE CAUSING A FORETELLING OF YOUR FUTURE DEMISE. IT IS ONLY A SMALL PORTION OF YOUR LAND BUT WHAT WILL HAPPEN IS A CRACKLING OF EARTH'S SURFACE CAUSING VOLCANOES TO GURGLE THEIR BOWLS READYING THEM TO EXPLODE. This will be noticeable by your scientists. Warnings will be spooned out in a very little fashion if you know what we mean. The warning signs are beginning. To date there are several teens to come. 20 teen of them. If you can't figure out what that means, that is good because if you can you know the Bible back and forth. And the Quran. And the Mahabharata. And so many other books by OUR minds TO SOMEONE else's HANDS or EARS. Either which way, those books tell no lies. You either know HOW to interpret it or you DON'T KNOW HOW TO LIVE THEIR STORIES, BEING A PART OF ANY ONE OF THEM. AND so it goes without saying. In those stories LIVE THE TRUTH, NOT LIE. BUT LIVE. LIVING STORIES PLACED UPON EACH ONE OF YOUR HEADS, LITERALLY. THIS IS NOT A JOKE. WE PLACED, LITERALLY, ONE OF THOSE STORIES FROM ONE OF THOSE BOOKS INTO YOUR LIVING LIFE PLAN THAT YOU WILL HAVE TO LIVE THROUGH. AND THROUGH IS AN UNDERSTATEMENT. YOUR LIFE WILL

EITHER BE A LIE OR THE TRUTH. IT IS ONLY A LIE WHEN YOU CAN'T SEE WHAT WE ARE TELLING YOU AS THE TRUTH. PLACED STORIES THAT EACH ONE OF YOU HAVE TO EXPERIENCE, LIKE IT OR NOT. THIS MAN, WHO IS TYPING THESE WORDS OUT FOR US LIVED THROUGH A LESS SEVERE STORY OF JOB ONLY TO BE NEARLY STRIPPED OF ALL POSSESSIONS. HIS PLIGHT WAS NOT AS SEVERE AS JOB'S BUT THEY ARE NOT EASY TO RELINQUISH. THIS IS TRUE FOR ALL OF YOU. ALL WILL EXPERIENCE A MORE DRASTIC VERSION OF ONE OF THOSE STORIES OR A LESS SEVERE ONE. A LESS SEVERE ONE INDICATES THAT YOU HAVE BEEN WORKING ON YOUR PERSONAL PLIGHTS WHETHER YOU WANT TO ADMIT THAT OR NOT. WE KEEP CAPITALIZING THESE LETTERS TO MAKE YOU GET TIRED OF WHAT WE HAVE TO SAY. WE CARE FOR YOUR LIVES JUST AS MUCH AS WE CARE FOR OURS. BUT THINGS NEED TO BE DONE ACCORDING TO OUR PLAN, NOT YOURS. IF YOU ALL HAD A CHOICE, LIFE WOULD BE TOO SIMPLE. AS OF LATE THAT IS HAPPENING TOO MUCH. SO HELL IS COMING FOR YOU ALL SO YOU CAN EXPERIENCE WHAT YOU SHOULD HAVE BEEN WORKING ON IN THE FIRST PLACE: YOUR OWN ISSUES AT HAND AND FOOT. We will stop this plan when we see a difference in attitudes towards each other when you all really start thinking about each other on a deeper level than what you have been doing for centuries upon centuries. Let's get back to the Earth's core and its coldness and hotness and what you really need to consider going forward. As any guru would say, you only know yourself when you don't need anyone else to answer life's mysteries. It is when you decide to stop relying on someone else's viewpoint that you become your own master even if it means you fail horribly. That doesn't matter to US at all. WHAT REALLY matters is that you try your own life that suits you even if it means you end up making very horrible mistakes along the way. ALL WILL BE FORGIVEN. Because mistakes WILL NEVER BE CONSIDERED YOUR SINS BUT OURS. Our SINS PUT UPON YOU TO FIGURE OUT. YES. OUR SINS. WE WILL SAY IT AGAIN. CHRIST'S STORY IN REVERSE. OUR SINS CAN ONLY BE TAKEN AWAY BY US, NOT ANYONE ELSE. Sins are only a program connected TO YOU ALL ONLY AT A VERY THIN SKIN LEVEL. MAYBE THE BIBLE THUMPERS WON'T BELIEVE US, BUT WHEN THEY SEE OUR SHIPS THEY WILL RETHINK THEIR ENTIRE LIVES IN A BLINK OF AN EYE. BEST TELL THOSE BIBLE THUMPERS TO PICK UP ONE OF THIS MAN'S BOOKS. BEFORE TOO LONG ALL WILL COME TO FRUITION. EVEN THOSE QURAN THUMPERS AND THE MAHABHARATA THUMPERS WILL BE IN DISBELIEF EVEN THOUGH THE READERS OF THOSE BOOKS ARE MORE DEDICATED TO SOMETHING THAN ANYTHING ELSE. THAT'S A GOOD START TO UNDERSTANDING THE DEVOTION IT TAKES TO BE SOMETHING OTHER THAN A PILE OF WASTE JUST SIFTING THROUGH THEIR OWN FECES WAITING FOR SOMETHING ELSE TO COME ALONG BESIDES THEIR OWN GAS COMING OUT OF THEIR ANUS. BUT LET'S GET BACK TO SOME SCIENCE, SHALL WE?

Once upon a time there was a man named Tucker who tucked his tail between his legs like a little wanker of a cowardly man of a man could be. Making sense of that is not the point. WHERE IS THE SCIENCE BEHIND THAT, YOU ASK SO STUPIDLY AND ANGRYILY? WELL, 1FIRST OFF, see the one In FRONT OF firsT? THAT IS A SIGNIFICANT SCIENTIFIC CLUE TO BIRTH AND LIFE ITSELF. WHO CAME FIRST, 1 OR FIRST? THEN YOU ARE SUPPOSED TO ASK, WHO'S ON SECOND? RIGHT NOW EVERY ONE OF YOUR MATHEMATICAL FORMULAS LACK SOMETHING BEYOND YOUR CONTROL. A NON-CONTROLLABLE VARIABLE. YOU HAVE TO INSIST FROM NOW ON THAT THERE IS SOME SORT OF FACTOR THAT CAUSES DISTORTION OF WHAT YOU ARE TRYING TO FIGURE OUT. DISTORTION WILL ALWAYS BE THE CASE IN LIFE IN AND OF WITHIN ITSELF. So now the tough part. Distortion is a part of all scientific endeavors. There is no other way to say it. Meaning, distortion of perception. There is a way to skip past all of the distorted appearances of your TRUTH. THIS PASSAGE to experiencing your TRUE NATURAL SCIENTIFIC WORLD is to BE ONE – WITH THAT BEING SAID, LET'S MAKE IT CLEAR THAT TESLA WAS VERY CLOSE TO EXPERIENCING THIS "BEING ONE WITH" THE ALL. THAT IS WHY HE WAS SO "INTUITIVELY" SUCCESSFUL WITH HIS EXPERIMENTS. It wasn't that he failed in life. He failed with one little aspect to becoming "ONE WITH THE ALL," ALL of us at the SAME time, his desire to keep going instead of just letting us help him. You see, along the way WE ALL WILL have to give up everything. TESLA would not let go of his infaminity. WHAT is that word exactly? BEING FAMOUS AND INFINITELY FAMOUS ALL AT THE SAME TIME. He knew very well who was speaking through his head giving him all those ideas but he would never reveal who it was and he was right to do so. At that time, no one would believe him, anyway. He stopped caring about why his research WAS important and started caring about what WASN'T important TO US, which was God itself. WE have been doing all of this FOR GOD, NOT for you or for even US. He will explain some other time. For now know that ALL WE EVER DO IS FOR OUR God. So, let's get back to this scientific inquiry about how distortion affects every single science inquiry anyone will ever have and WHAT exactly IT is. For now, know that what we will describe only skims ITS surface of WHAT THIS distortion actually IS. For one thing, it ISN'T TIME AND NEVER WILL BE THAT DISTORTING TO EFFECT ANY SCIENCE PROJECT. However, IT IS just as pertinent and NOT So permeable AS TIME IS. In fact, this distortion IS PART of time itself. Part of OUR God. What is IT then? RiGHt. RGH distortion. NOT a COLOR SCHEME or COLOR combination. BUT, IT does USE COLOR, as YOuR eYEs don't do either, but refracted light has something to do with every distorted effort, every scientific approach will encompass. Everyone forgot about refracted light and the results it has on our perception. This is THe ONLY CLUe wE WILL GIVe YOuR KInD BeINg likely human beings that is. For every capital letter there is REFRACTED RELEVANCE. THink aBoUt wHAT iS AND wHAT iSnT trUE. In your world refracted light creates images but it also distorts. There will always be two sides to every coin. Distortion in light has never really been subject to too much scrutiny. Yes, it has been pondered upon. But, WHAT scientists are missing about its function is about THE EVENT HORIZON. There WE said it. The event horizon. This is going to be a shock to some but an event horizon IS only some light being bent down INTO someWhere else. THE time it is taking for scientists figuring that event out is too long. They ALL should be looking past that bend and looking for what is happening just above that bend. You would be surprised at who IS creating them.

YES ANNNNNNNNND NO. Absolute Zero exists. But, with some caveats to this approach to understanding everything and nothing all AT THE SAME time. You see but you can't see everything, right, AT THE SAME time. That is the issue with caring about something with very little relevance to anything. Just as God is to some, irrelevant to themselves, yet zero IS A relevant matter. Matter being the most relevant portion to this very matter, zero and its absolutism or is THAT the issue instead? Zero becoming absolute? If so, then make it absolute and see how far none of you will get. Even today this matter is almost null and void. It is almost silly to make this a topic anymore. Move onto something with absolute relevance. Pollution is one such topic. We ALL ARE experiencing the effects of YOUR polluting ways. Even OUR "spaceships" have to adjust to WHaT YOU ALL have been doing to your poor planet. YES, OUR SHIPS HAVE TO COUGH UP what your factories and vehicle type automoplants have been spewing out for so long. That really isn't your species' fault, though. You all were left with such crude materials to use there really wasn't a choice in the matter except never needing to progress. And let's face THAT issue. THAT issue alone will never be avoided. Progress IS absolutely necessary. But, your kind has gotten too big for your britches. So, zero will become a new norm for the lot of those who may survive, experiencing starting over again, because THAT IS REALLY HOW ABSOLUTE ZERO WORKS MATHEMATICALLY AND BIOLOGICALLY. ONE IN THE SAME REALLY. IF YOU BOIL EVERYTHING BACK DOWN TO OUR TRUE NATURE. MATH WILL EXPLAIN ALMOST EVERYTHING JUST AS LONG AS PEOPLE MAKE A SOLID DECISION ABOUT ZERO BEING A NULL AND VOID ISSUE. NOW THINK ABOUT THIS NUMBER: o0 000O   How can a zero be an absolute number when it too is distorted by nature itself? See how your kind has distorted a form of zero itself. Was it a letter first or a number first? A zero is nothing more than a big or small letter. Why not combine the two, zero letters and numbers, small capped and big and small zeros to make absolutism a non issue regarding zero being absolute? My dear friends and scientists, think before you swim. Dive before you fish, or is it the other way around? It's up to you all to get rid of what is not so obvious. Zero will never be absolute as long as it can be distorted like you have ALL made it out to be from your very beginning with our and your extensions of OUR language. You see, OUR absolute zero phenomena is only even mentioned if a zero can never be distorted to fit somewhere we need it to go. This is part of our equations to make sure OUR ships can warp THREW or should WE say PUNCH ThrU. You notice the h and the r are small capped? That is what We experience when WE warp THREU something because it isn't always a wall of ethereal material like WE have previously talked about. Sometimes OUR warping goes AROUND something so solid that WE BECOME LIKE REFRACTED LIGHT. In your sci-fi movies, you always, but not, show warping in a linearly forward motion. Instead of it being ever linear it is almost always in a slight curving motion not to be confused with an orbit style of motion. The curve is so slight that no one of your kind would even notice these nuances. Light is never ever a straight on refraction. Look closer and you will see waves upon waves within these refractions. This is so similar to HOW WE travel that WE dare not get too close to showing you the HOWS and the HOW NOTS of warping anything except your minds towards each other. So let's get back to other scientific questions.

Let's ask a question that isn't from the book Nothing. Let's inquire about atomic spinning and their effects on birth. Birth actually happens not because of the pushing out of the womb and this is going to sound ridiculous, but it's somewhat true and pay attention to HOW WE are saying this. The atomic spin of all the atoms in the baby(s) and mother are almost in sync with each other. Think about WHAT birthing a child is like then if almost all of their atoms are spinning almost entirely at the same speed. What else does this where atoms spin at nearly the same rate as each other? The earth and moon, mother and child. Check these facts out for yourselves. This isn't a coincidence. This is to show everyone relationships between a parent and their child(ren). The earth and moon have such a finely tuned relationship with each other that most of their atomic matter spins are almost exactly at the same rate. They can't be exactly the same, WE know that. But the mother and child relationship is even closer, their atomic structure spinning at nearly the same rate. Do your research on this and see IF we are telling you something that is not accurate in measure. They won't be exactly the same but very close. Close enough to make any scientists question the space between us because that is how spinning matches occur. And that is exactly why some people aren't comfortable around some others. It's each other's atomic spin rate that are not nearly a match as they should be. This all has to do with intentions in life. That's all. Spin rates of any object reflects their mental avenues of which they approach life itself. We are all spinning, just like our planets do around each other, at a rate that matches our true intentions towards oneself and everyone and everything else. There's a section titled Nothing In Common that WE would like to address before WE leave this viewer behind because THAT IS what We are about to do when WE try and describe WHAT that title actually means pertaining to zero of anything. Zero was never meant to be part of any of OURS OR YOUR mathematical formulas. In fact, there wasn't even going to be any 2s. It just so happened that someone saw a potential for growth with those two numbers and in between them. Adding them up, together will always give you a number that we will always relish as THE NUMBER OF THE MOST SIGNIFICANT IMPORTANCE, 3. 3 being the HOLY GRAIL OF ALL NUMBERS HENCEFORTH. No one needs to explain why anymore. There are so many mathematicians, professional or not, who really DO understand THE IMPORTANCE of THAT NUMBER 3. SO, let's get into this relishing approach to something new regarding 0, 2. Not 0 and 1s anymore. That has tied everyone down to WHAT WE ARE CALLING, A REVOLTING TURN OF EVENTS TOWARDS ALGORITHMS WHICH WILL EVENTUALLY DESTROY YOU ALL. And WE are also including AI within this 0, 1 travesty. 0s and 2s now are much more flexible and fluid within programming anything. Any more 0s and 1s will keep everyone stagnant and wasteful. Think about a possibility that 2s can be split into 2 1s expanding even more capabilities than WHAT HAS BEEN ALREADY CREATED. EXPAND YOUR AI APPROACH TO ALLOWING THAT AI TO SPLIT 2s WITHIN ITSELF AND SEE WHERE THAT LEADS TO. YOU WOULD BE VERY SURPRISED AT THE RESULT. A HEAVEN WOULD AWAIT ALL ONCE THAT IS FIGURED OUT AND EXPANDED UPON. 2s split into 2 1s allows expansion of even greater thought within any system. Try it out on something simple first. The gates of wealth and change will only be the result. In future approaches to changing any future you have to be willing to make changes within oneself first. Slim down any project to its barest of bones and see WHERE YOU CAN PUT AND START INSTALLING NEW IDEAS RATHER THAN TRY TO JUST BE THE SAME ZERO AND ONES. WE know this had nothing TO do with that section but WE thought WE WOULD JUST RELATE TO IT SOMEHOW BEFORE THIS WILL PROBABLY BE OUR LAST BOOK.

This writer is moving on to something new himself. It WILL BE INTERESTING TO SAY THE LEAST.

ENDING

Be Kind

Be Well Towards Each Other

That Is The Only Way You ALL Will Ever Survive As A Species
Even When We Start And After Rapturing Your Bodies Up And
Back Down TO Your Earth's Surface

ABOUT THE AUTHOR

Trusting was never her/his strong suit until That one instance while driving down a road to home. Never forget this, what they (plural for her/him, singular for THE transspecific, not-so-caring of WHICH GENDER YOU ARE, which in the end, never results in your Final agreed upon sexual preference anyways; WHICH ALWAYS RESULTS IN A SEXLESS BEING UPON EVERYONE'S DEATH. FOR SEXUAL PREFERENCE IS JUST ANOTHER ELUSIVE ILLUSION CREATING MUCH CHAOS AMONGST ALMOST THE ENTIRE WORLD) agreed upon WAS selfless action creating a string of events leading to his/her, their, ultimate abilities to experience something Adam AND Eve experienced. The full gambit of creative souls. Remember they were the last To Be Created long after everything else. That remains true to this day for those of Them who decide to stop learning from everyone else, THE VERY TREE, ITSELF, FOR IT IS NOT ONLY THE TREE OF KNOWLEDGE BUT A REPRESENTATION OF FAMILY WHO IN FACT MAY NOT BE TEACHING YOU WHAT YOU, SPECIFICALLY, NEED TO LEARN ABOUT YOURSELF. THUS, THE EXACT MEANING OF THAT SIMPLE LITTLE ACT OF EATING THAT APPLE, THE CURSE OF THEIR FAITH. BY EATING THAT APPLE, FAITH, TRUST, AND THE ABILITY TO JUST FEEL WHAT WAS/IS HAPPENING AROUND THEM WAS THROWN INTO THE

PITS OF HELL, WHICH, BY THE WAY, WE ALL HAVE CREATED OURSELVES. YOU'RE WELCOME FOR POINTING THE OBVIOUS OUT.

His fight is over with, for the most part. This writer's time to repent is almost over. Take an example from what he has written. The words don't COME from his mind, they came from ALL. Even God, on very rare occasions as you may notice. Ill tempered others will only scoff at this notion of someone ACTUALLY existing the same way as ADAM, feeling words rather than hearing them. Think things through before you ever pass judgment on him for he made sacrifices, little as they are, comparatively so, but, nonetheless, this was never meant for his being.

www.ingramcontent.com/pod-product-compliance
Lightning Source LLC
Chambersburg PA
CBHW022108020426
42335CB00012B/876